LETTS POCKET GUIDE TO

MUSHROOMS
& Other Fungi

The most familiar species of European
mushrooms and fungi described and
illustrated in colour

Eleanor Lawrence and Sue Harniess

Warning Symbol

 **Throughout the book this sign has been used
to indicate that a mushroom is toxic,
indigestible or has a highly disagreeable taste,
and should never be eaten. Some mushrooms
are deadly and therefore the reader is advised
to take great care in identification of all types.**

If in any doubt DO NOT EAT!

Always consult an expert!

Front cover illustration: Spring Bolete (right) is good to eat, whereas
Satan's Bolete (left) is **toxic**

This edition first published 1990
by Charles Letts & Co Ltd
Diary House, Borough Road,
London SE1 1DW

'Letts' is a registered trademark of
Charles Letts (Scotland) Ltd

This edition produced under licence by
Malcolm Saunders Publishing Ltd, London

ISBN 1 85238 102 7

Printed in Spain

Contents

Introduction

Fungi are amongst the most fascinating members of the vegetable kingdom. Traditionally regarded as plants, they are now placed by biologists in a kingdom of their own. Unlike green plants they contain no chlorophyll and cannot make their own food using the sun's energy. Instead they extract nourishment from the soil or from the wood or other material on which they live, causing it slowly to rot away. They are a most important part of the great cycle of life, releasing elements locked in dead animal and plant matter to be returned to the soil and used again.

There are many thousands of species of fungi, some so small they are only visible under the microscope. In this book we present a selection of the larger fungi, including many that are gathered for food and those deadly poisonous species that must be avoided at all costs.

Identifying fungi for the table requires great care. If you are a complete novice do not rely entirely on books but also get your finds identified by someone knowledgeable. A good way to start is to go on an organized mushroom hunt with a local mycological society or mushroom club.

Always check very carefully that a mushroom you think is edible corresponds in *all* respects to its description in the book. A difference in spore colour, for example, could mean that you have gathered a poisonous species. **If you are at all unsure, do not eat it!** Before you go out collecting, look through the whole book and familiarize yourself with the better-known edible and poisonous species and note the ones that can be confused. Always gather the complete mushroom, including the very base of the stalk, so that you have all the parts needed to make an identification. Where possible also gather several of the same species at different stages in development. Collect mushrooms in a wide, flat-bottomed basket, not in plastic bags in which they become easily damaged and identifying features lost. Finally, when trying a new species for the first time, only eat a very little of it, as even edible mushrooms can cause illness and allergies in some susceptible people. Always keep one of the mushrooms in case it has to be identified by experts later.

All the mushrooms noted as edible in this book are common species that have been eaten for many years. Those not designated as either edible or poisonous should be regarded as inedible.

Very few mushrooms have familiar common names, and common names also vary widely from place to place and from book to book. Since it is so important to be sure exactly which mushroom one is dealing with the Latin name has also been included in each case. Although these may seem offputting at first it is well worth getting to know them as it is then much easier to check your finds with an expert or with other books.

How to use this book

The book has been divided into five main sections indicating the type of habitat in which certain mushrooms are most likely to be found. Each habitat is indicated by a different colour band at the top of the page (see Contents page 7.) Some mushrooms grow in a close relationship with the roots of certain trees and are only found in association with them. Others, although they prefer a certain type of habitat, may also be found elsewhere. The five sections are:

Broadleaved Woodland
Most commonly found growing on the ground in broadleaved woodland or in association with broadleaved trees such as beech, birch and oak.

Coniferous Woodland
Most commonly found growing on the ground in coniferous woods or forests or in association with conifers such as pine and spruce.

Mixed Woodland
May be found growing on the ground in mixed woodland or in association with either coniferous or broadleaved trees.

Grasslands and Parks
Most typically found in grass or open situations, as in lawns, pasture, parks, gardens, roadsides, woodland glades, waste places, heaths, moors.

Growing on Wood
Growing directly on wood, as on living trees, stumps, cut timber, rotting wood, twigs and branches on the ground.

Within each main section fungi have been placed into subsections possessing certain identification features in common. These groups have been chosen to identify the species illustrated in this book only and do not necessarily reflect biological relationships. Within each colour band these subsections are identified by symbols.

Identification features
Mushrooms are the fruiting bodies of certain sorts of fungi. The fungus lives for most of the year as a mass of thin threads known as a *mycelium*, which is sometimes visible as a cottony mass at the very base of the stalk. At the appropriate time of year, in late summer and autumn for most mushrooms, this mycelium produces fruiting bodies containing spores by which the fungus propagates itself.

In some mushrooms the spores are borne on thin leaf-like *gills* on the underside of the cap (Fig. 1) in others (boletes, polypores and relatives) they are formed in close-packed tubes that take the place of the gills.

Important features to look for are the shape of the cap (Fig. 2), whether the mushroom has a *ring* or *volva* on the stalk (see Fig. 1), and how the gills are attached to the stalk (Fig. 3). Not all mushrooms have all the features shown. Many mushrooms do not, for example, have a ring or a volva. In some mushrooms, such as cortinarias (corts), the young gills are covered by a cobwebby veil, the *cortina*, which sometimes leaves traces on stalk and edge of cap. Always handle mushrooms carefully when gathering them to preserve the surface texture of the cap and stalk which are also useful in identification.

Fig. 1 *Left* Section through a gilled mushroom showing parts useful for identification. *Right* Immature mushroom showing how ring and volva are formed.

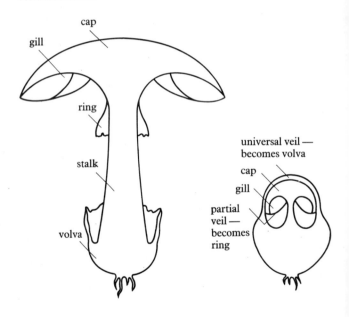

Fig. 2 Types of cap shape

Convex   Hemispherical   Umbonate   Conical   Vase-shaped

Fig. 3 Types of gill attachment

| Free | Adnexed | Adnate | Decurrent | Sinuate |

Taking a spore print

The colour of the spores is also an important identifying feature, helping to divide mushrooms into several large groups, and to distinguish between rather similar mushrooms belonging to different groups. To take a spore print the mushroom is cut off just below the cap and placed on a sheet of white paper (or black paper if you suspect you will get a white spore print). A glass or a plastic bag is placed over it. After a period ranging from a few hours to a day, spores will drop from the gills or pores to make a print on the paper.

Identification symbols

Within each colour band fungi with certain features in common have been grouped together, and each group is identified by the following symbol at the top of the page.

Fungi with teeth or ridges on undersurface, not gills or pores

e.g. Horn of Plenty, Chanterelle, Bear's Head Tooth.

Fungi with pores on underside of a cap or a bracket

Central stem and rounded cap, growing on ground. These are the boletes.

Fungi with very short or no stalk, or shelf-like, growing on wood, e.g. polypores, Turkey-tail.

Fungi with gills on underside of a cap or bracket

No or very short stalk, or shelf-like, growing on wood, e.g. Oyster Mushroom.

Typical mushroom-like fungi with gills on the underside of a cap and a central stalk

 Small fragile mushrooms, generally no more than 6 cm:2½ in high with a delicate cap. Cap generally convex, hemispherical, or conical, e.g. mycenas, psilocybes.

 More robust mushrooms with gills that reach the stem (i.e. attached gills), and no ring or volva, e.g. russulas, trichs, milk-caps, and many others.

 More robust mushrooms with attached gills, no volva, possessing a cobwebby veil (the cortina) over gills in young specimens, sometimes leaving a ring-like mark, or series of marks, on stalk, e.g. corts.

 More robust mushrooms with ring on stalk, no volva, and attached gills, e.g. agrocybes, Honey Fungus, Gypsy.

 More robust mushrooms with no ring or volva on stem, gills do not reach stalk (free gills), e.g. Fawn Mushroom.

 Larger mushrooms with no ring on stem but which have a volva and free gills, e.g. volvariellas.

 Larger mushrooms with free gills, a ring on stalk, but no volva or with volva reduced to scales at base of stalk, e.g. agarics, some amanitas, lepiotas.

 Larger mushrooms with free gills and both ring and volva, e.g. some amanitas.

Fungi without teeth, gills or pores, fruitbodies not in
 typical mushroom shape, e.g. stinkhorn, morels, puffballs, earthstars, earth-balls, coral-fungi

After having determined, with the help of the colour bands and the symbols, the section in which your mushroom is likely to be, look at the illustrated pages giving detailed descriptions of each species. The size is indicated on the coloured band by measurements for cap diameter and length of stalk.

Four boxes provide information for identification. The first gives features which together with the illustration should enable you to distinguish that species from others illustrated in this book and wherever possible from others which you might come across, especially poisonous species. The second box gives supplementary information to help identification, and notes on edibility or otherwise.

Distribution and habitat of the illustrated mushrooms are given in the third box, and the fourth box gives species with which they might be confused. Those in **bold type** are illustrated elsewhere in this book.

There are more than a thousand species of larger fungi in Europe, and so you will inevitably find fungi not illustrated in this book. Nevertheless, you will be able to identify some of them to their family group or *genus*, such as *Agaricus*, *Russula* or *Lactarius*, by noting the spore colour, type of attachment of the gills etc., and looking for similar species in these pages. However, **do not eat anything you have not identified precisely**, as a single genus can contain both edible and poisonous species.

Specimen page

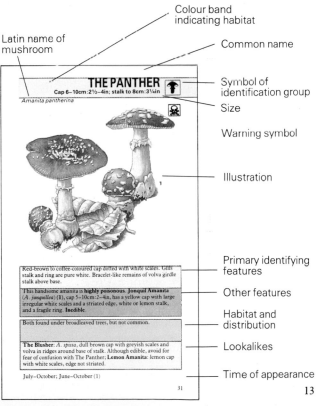

Colour band indicating habitat

Latin name of mushroom

Common name

Symbol of identification group

Size

Warning symbol

Illustration

Primary identifying features

Other features

Habitat and distribution

Lookalikes

Time of appearance

THE PANTHER
Cap 6–10cm:2½–4in; stalk to 8cm:3¼in
Amanita pantherina

Red-brown to coffee-coloured cap dotted with white scales. Gills stalk and ring are pure white. Bracelet-like remains of volva girdle stalk above base.

This handsome amanita is **highly poisonous**. **Jonquil Amanita** (*A. junquillea*) (1), cap 5–10cm:2–4in, has a yellow cap with large irregular white scales and a striated edge, white or lemon stalk, and a fragile ring. Inedible.

Both found under broadleaved trees, but not common.

The Blusher: *A. spissa*, dull brown cap with greyish scales and volva in ridges around base of stalk. Although edible, avoid for fear of confusion with The Panther; **Lemon Amanita**: lemon cap with white scales, edge not striated.

July–October; June–October (1)

31

Craterellus cornucopioides

Blackish or dark grey-brown, narrow funnel-shaped cap with a flared wavy lip. Outer surface of cap is smooth or faintly ridged, the inner surface slightly scaly or velvety.

A small to medium-sized vase-shaped fungus with a short hollow stem and thin greyish flesh. Spores white. One form has a yellow outer surface. Good to eat and can be dried and stored.

Under broadleaved trees, especially oak and beech, often in clusters. Relatively common throughout Europe.

The similar *Cantharellus cinereus* has well-defined forked ridges on outer surface. Although no vase-shaped fungus with ridges on underside is toxic, take great care not to confuse them with similar-shaped species with true gills.

August–November

14

Boletus reticulatus (aestivalis)

1

Uniform pale brown to brown matt surface to cap, and pores remaining whitish for a long time. Stem reddish brown covered all over with a distinct fine white network.

A typical bolete, with a fleshy cap whose surface tends to crack, and thick stem swelling towards the base. The colour of the pores remains unchanged when bruised. The flesh is white and firm. Good to eat, with a sweet nutty taste.

Under broadleaved trees, especially beech. Appearing earlier in the year than the Cep.

Cep and other boletes; **Bitter Bolete**, **inedible**; **Satan's Bolete** (*B. satanas*) **(1)**: **toxic**, pale grey cap, red pores, bright red network on a stem swollen at base; *B. fragrans*: dark brown cap, strong fruity scent, **edibility doubtful**.

May–October

Krombholziella scabra

Rough stalk covered with hard small brown or black scales. Cap greyish brown to dark brown, pores off-white to buff. Flesh unchanging when cut, or bruising slightly brownish.

A tall-stalked bolete with a cap dry, greasy or sticky to the touch, starting convex and sometimes becoming flatter with a depression in centre. Pores change from pale buff to greyish with age. Edible and good.

Under broadleaved trees, especially birch. Widely distributed and quite common throughout Europe.

Several scaly-stalked boletes grow in association with broadleaved trees. They differ in cap colour and texture and in the discoloration of the flesh on cutting. None is poisonous, the **Red-capped Bolete** being considered the best to eat.

June/July–November

Xerocomus chrysenteron

1

The dull olive-brown skin of the cap becomes cracked, showing a pink layer underneath. Pores dirty yellow to greenish. Flesh yellow, sometimes turning blue or green on bruising.

Medium-sized bolete, with stalk slightly bent towards the foot. The **Yellow-cracked Bolete** (*X. subtomentosus*) (**1**) has a velvety pale buff to olive-brown cap, becoming cracked, golden pores and a cream stalk. Both edible.

Under broadleaved trees, in grassy copses. Both are common throughout most of Europe.

Other *Xerocomus* species occur under broadleaved trees, with brown or red caps, sometimes cracking, but no others show pink flesh underneath. One species is parasitic on earthballs. The **Bay Bolete** grows under conifers.

August–November

CHESTNUT BOLETE
Cap 5–10cm:2–4in; stalk to 10cm:4in

Gyroporus castaneus

Small bolete with rusty brown to chestnut cap, white or pale yellow pores, not changing colour when bruised. Stalk same colour or paler than cap, rather uneven, brittle and hollow.

The surface of the rounded fleshy cap is smooth or slightly velvety and the stalk also has a slight bloom when young. The white flesh does not change colour when cut. **Edibility suspect**.

Under broadleaved trees, especially oak and beech, often growing in clusters. Quite common and widespread.

Bluing Bolete; **Ceps** and other **boletes**; **Slippery Jack** and other *Suillus* spp.

August–October

GREEN-CRACKED RUSSULA

Cap 10–15cm:4–6in; stalk to 5cm:2in

Russula virescens

Brittle gills and flesh typical of a russula. Pale green or bluish green matt cap surface flaking and cracking to show white underneath. Creamy white gills. White spores.

Flesh hard and white, spongy in the stalk. Edible, with a mild nutty taste. **Forked-gill Russula** (*R. cyanoxantha*) (**1**) cap is bluish green to violet, not cracking. The gills are distinctly forked. Edible, good.

In broadleaved woods, especially beech. Both are widespread and common.

Green Russula: gills forked but more crowded; *R. cutefracta* has a darker green or bronze-tinted cap, cracking inwards from the edge. Edible, although some forms may be acrid.

July/August–November

19

Green Russula (*R. heterophylla*)
(**1**) has a smooth, shiny green to
bronze cap, 8–15cm:3¼–6in
across, crowded white gills,
forked. Spore print white.
Edible, but take care to
distinguish from the deadly
Death Cap.

Bare-toothed Russula (*R. vesca*)
(**2**) has a pinky-brown cap, 5–
9cm:2–3½in across, becoming
dull yellow in centre. Skin peels
off easily and mature specimens
show white flesh at edge of cap
where skin has retracted. Spore
print white.

Golden Russula (*R. aurea*) (**3**) is
one of many bright red-capped
russulas found in broadleaved
and mixed woods. It has a brick-
red cap and bright yellow gills
and flesh. It prefers warmer
areas. Spore print ochre to
yellow.

MILK-CAPS (LACTARIUS)

The many different species of Milk-cap look similar to russulas but the gills release a milky fluid when broken. All have a white spore print. Many are hot and peppery to the taste, but a few are edible.

Slimy Milk-cap (*L. blennius*) (**1**) cap 6–10cm:2¼–4in across, olive-brown to grey-green, often with concentric rings of drop-like marks around edge, and becoming sticky and slippery when wet. Milk turns grey on exposure to air. Flesh very acrid. Under beech.

Sweet Milk-cap (*L. subdulcis*) (**2**) cap 3–6cm:1¼–2½ in across, pinkish tan, with a surface like suede leather. Gills pale cream or tinged pink, stalk cream at top becoming brown lower down. The milk stays white. Mild taste. Edible. Common under beech and other broadleaved trees.

Oak Milk-cap (*L. quietus*) (**3**) cap 6–10cm:2½–4in, dull reddish brown with darker zones. Gills creamy. Milk white. Flesh has a smell sometimes described as "like wet linen". Under oak. **Inedible**.

PEPPERY MILK-CAP
Cap 6–12cm:2½–4¾in; stalk to 8cm:3¼in

Lactarius piperatus

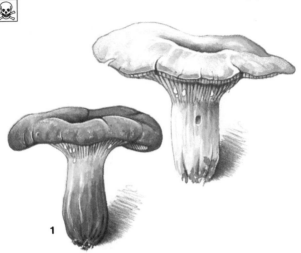

White cap becoming sunken in centre. Cream or slightly pinkish crowded gills exude a copious, very acrid milk, which stays white. White stalk does not discolour on bruising.

Cap has a matt, wrinkled surface and edge rolled under. The gills run down the stalk slightly. **Voluminous-latex Milk-cap** (*L volemus*) (1) has a pale orange, matt, wrinkled cap, crowded yellow gills and smells fishy when handled. Edible.

Peppery Milk-cap: broadleaved copses on neutral or limy soil. Voluminous-latex Milk-cap: especially under oak.

Other white milk-caps include *L. vellereus* which is larger (12–35cm:4¾–14in across) with a lobed edge to cap and short stout stalk; other white or grey milk-caps have milk changing colour or shorter stalks. None is edible.

August–November

WOOLLY MILK-CAP
Cap 8–12cm:3¼–4¾in; stalk to 8cm:3¼in

Lactarius torminosus

1

Cap pinkish to flesh-coloured, with concentric paler zones. The incurved edge is very shaggy or woolly. Pinkish gills release a white unchanging acrid milk. Spore print cream.

White flesh smells faintly of geranium leaves. Stem white, pitted pink. **Inedible, considered poisonous.** *L. chrysorrheus* (1) has a smooth cap with a yellowish tinge and gills and stem bruise yellow. Acrid but edible after cooking.

Woolly Milk-cap: under birch, often in mixed woods and on heaths, common. *L. chrysorrheus*: common under broadleaved trees, especially oak.

Saffron Milk-cap: in conifer woods.

August–November

Clitopilus prunulus

White velvety cap often becoming sunken in centre, with lobed incurved edge, and pink gills descending a white stalk. Strong smell of new-ground wheatmeal. Spore print pink.

The lobed edge to cap is more marked in older specimens. Gills are cream at first. The thick white stalk is sometimes placed off-centre. A good edible fungus but may be confused with several poisonous species (see below).

Common and widespread under various broadleaved trees, often on heaths or in grassland.

Sweating Mushroom, poisonous; **Lead Poisoner**, poisonous; **Field Mushroom** and other *Agaricus* spp, ring on stalk, gills free; **St George's Mushroom**.

July–November

DEADLY INOCYBE

Cap 3–7cm:1¼–2¾in; stalk to 8cm:3¼in

Inocybe patouillardii

Silky whitish conical to bell-shaped cap with a lobed edge becomes a dull orange with age. Gills, flesh and cap surface redden on bruising. Stout white stalk. Spore print brown.

The cap edge often splits towards the centre as it expands. This mushroom is **deadly poisonous**. It has a fruity smell which becomes earthy.

At the edges of broadleaved woods and under broadleaved trees in parks and gardens.

I. adequata: silky dark brown, twisted fibrous stalk; *I. rimosa*: fawn with darker silky fibrils, gills yellowish green; *I. maculata*: nut-brown with white patches at centre of cap, gills beige. All spore prints brown.

May–November

SOAP-SCENTED TRICHOLOMA
Cap 10–15cm:4–6in; stalk to 12cm:4¾in

Tricholoma saponaceum

1

Fleshy fungus with a dark grey to grey-brown cap, gills far apart, sinuate (typical of all tricholomas), white or pale yellow. Distinctive soapy smell. Spore print white.

Cap smooth and silky, cracking into scales in dry weather. Flesh pale. **Inedible. Sulphur Tricholoma** (*T. sulphureum*) (**1**) cap 4–8cm:1½–3¼in, is bright yellow with an unpleasant smell. Gills sinuate-adnexed. White spores. **Poisonous.**

Soap-scented Tricholoma: widespread and common under broadleaved trees, also under conifers. Sulphur Tricholoma: locally common, especially under beech, oak and holly.

Death Cap; Marvellous Tricholoma; White Tricholoma; Leopard Tricholoma; tricholomas are distinguished by their sinuate gills and white spores. **There are several poisonous species.** A few have a ring-like marking on the stalk.

August–November

LEAD POISONER

Cap 8–20cm:3¼–8in; stalk to 12cm:4¾in

Entoloma sinuatum (lividum)

One of the larger entolomas. Creamy yellow to pale coffee coloured slippery cap. Gills adnexed, almost free, not crowded, yellowish, turning pink. Silky stalk. Spores pink.

Large fleshy fungus with cap edge incurved in young specimens becoming wavy-edged when old. The thick white firm flesh smells slightly of cucumber or new-ground meal. **Highly poisonous**, causing cramps, severe vomiting and diarrhoea.

Widespread under broadleaved trees, also in parks, wasteland and by roadsides.

Cloudy Agaric; **St George's Mushroom**; **The Miller** (decurrent gills); other entolomas (sinuate or adnexed gills, pink spores); other tricholomas (sinuate gills, white spores).

July–November

27

Oudemansiella radicata

Tall mushroom with tough slender stalk ending in a long underground "taproot" eventually contacting wood. Sticky yellowish brown cap, gills pure white.

The cap often has wrinkles radiating out from centre, and thin pale flesh. Gills are thick, spaced far apart, and adnexed. Stem white to brownish, fibrous and twisted. Spore print white.

Common and widespread in woods, especially under beech and oak.

The rarer *O. longipes* has a dry, slightly hairy, brownish cap, no taproot, and long, twisted stalk covered in velvety brown hairs.

June–November

RED-BANDED CORTINARIUS

Cap 8–12cm:3¼–4¾in; stalk to 12cm:4¾in

Cortinarius armillatus

Large fleshy mushroom with reddish brown cap and stalk, the stalk bearing bright red irregular diagonal bands which are the remnants of a partial veil. Rust-brown spore print.

The bell-shaped cap ranges from fawn to dark russet, with a slightly silky surface. Stalk thick, pale russet with a bulbous base. Gills pale brown becoming dark rust. The white cortina leaves a whitish band on upper stalk.

Widespread and common under birch, on acid soil, especially near damp ground.

The many species of **Cortinarius** have variously coloured caps but all have rust-brown spores. Gills adnate or sinuate. **None should be eaten** as there are several deadly species,

August–October

29

 # GRISETTE
Cap 6–12cm:2½–4¾in; stalk to 12cm:4¾in

Amanita vaginata

1

Medium to large mushroom on a tall slender white ringless stem, the base enclosed in a white sac-like volva. Dull fawn or greyish cap has a striated edge. Gills and spores white.

This is edible after cooking to destroy toxins, but is thin-fleshed. The **Tawny Grisette** (*A. fulva*) (**1**) has a bright fawn cap and cream gills. Also edible after cooking. **Do not mistake deadly amanitas for these mushrooms.**

Both are common and widespread under beech, birch and oak, the Grisette also sometimes being found in coniferous woods.

A. batterae: close-fitting volva, yellowish-grey cap and stalk, rare, under conifers; **amanitas**: ring and sac-like volva or volva remnants at base of stalk.

June–October; May–October (1)

THE PANTHER

Cap 6–10cm:2½–4in; stalk to 8cm:3¼in

Amanita pantherina

Red-brown to coffee-coloured cap dotted with white scales. Gills stalk and ring are pure white. Bracelet-like remains of volva girdle stalk above base.

This handsome amanita is **highly poisonous. Jonquil Amanita** (*A. junquillea*) (**1**), cap 5–10cm:2–4in, has a yellow cap with large irregular white scales and a striated edge, white or lemon stalk, and a fragile ring. **Inedible.**

Both found under broadleaved trees, but not common.

The Blusher: *A. spissa*, dull brown cap with greyish scales and volva in ridges around base of stalk. Although edible, avoid for fear of confusion with The Panther; **Lemon Amanita**: lemon cap with white scales, edge not striated.

July–October; June–October (1)

CAESAR'S MUSHROOM
Cap 10–20cm:4–8in; stalk to 12cm:4¾in

Amanita caesarea

Smooth to greasy bright orange-red cap, occasionally bearing a fragment of the white universal veil. Stalk, gills and ring yellow. White sac-like volva at base of stem.

An excellent edible mushroom, highly esteemed by the ancient Romans. Flesh is yellow near surface. Spores white.

Under broadleaved trees in southern Europe no further north than the limits of vine cultivation. Not found in Britain.

The more northerly **Fly Agaric** has white patches on cap.

July–October

Amanita phalloides

Smooth greenish yellow to olive cap, sometimes almost white, sometimes bearing remnants of white veil. Gills white. Stalk bears a ring and large white volva. White spores.

This handsome mushroom is **deadly poisonous**, even in tiny amounts. It has a sweet smell rapidly becoming sickly. The young fruitbodies or "eggs" are completely enclosed in a universal white veil. Stalk pale, flushed same colour as cap.

Widespread and locally common. Mainly under broadleaved trees, especially beech and oak, but sometimes under conifers.

Other amanitas. All the deadly amanitas have white spores, white gills, and bear both ring and volva. **Wash your hands** thoroughly after handling any of them.

July–November

33

Amanita virosa

1

A shining pure white mushroom, occasionally with a pink tinge to cap, with a flaring fragile ring and large volva at base of stalk. White crowded gills. White spores.

The slightly greasy cap and scaly stalk distinguish this **deadly poisonous** amanita from the **equally deadly** all-white ***A. verna*** (1) which has a faint greenish tinge to the centre of the cap. **Wash hands** thoroughly after handling.

Both are rather uncommon, *A. virosa* being found under beech, birch and oak and also in mixed woods, *A. verna* being more common in mountains in southern Europe.

Young specimens can be confused with edible mushroom (*Agaricus*) species; **Death Cap** (white form); **avoid all-white mushrooms** as they include several deadly species; **White Tricholoma**: no ring or volva, white spores.

August–October

COMMON MOREL

Cap 3–7cm:1¼–2¾in high; stalk to 17cm:6¾in

Morchella esculenta

1

Rounded oval cap with a surface like a honeycomb, not separable from stalk at lower edge. Light yellowish brown to grey brown. Stalk whitish, soon becoming hollow.

One of the best known edible fungi which should, however, be eaten with caution as it disagrees with some people. The related Fluted **White Helvella** (*Helvella crispa*) (**1**) and similar fungi are not regarded as edible and are **toxic** raw.

Common Morel: sandy copses of broadleaved trees, also on rubbish heaps. Fluted White Helvella: damp broadleaved woods.

Forms of the common morel with conical caps are sometimes considered as separate species. **False Morel: poisonous**; *Mitrophora semilibera*: cap edge free from stalk, edible; *Verpa* species: rare, smooth cap 1–2cm:½–¾in across.

April–June; spring and August–October (1)

BLACK TRUFFLE
Fruitbody 6–12cm : 2½–4¾in diameter

Tuber melanosporum

1

Fruitbody grows completely underground. Shining knobbly coal-black outer surface, interior brown flesh marbled with white. Strong fragrance.

The most highly prized of all fungi. Specially trained dogs are used to search them out. The **White Truffle** (*T. blottii (aestivum)*) (**1**) is paler, and is covered with regular pyramidal warts. It is also edible but less good.

Black Truffle: uncommon, found in southern Europe, e.g. southern France and Italy, under oaks, in hilly areas. Not found in Britain. White Truffle: found also in Britain, under beech.

There are several other species of truffle, which are harmless, but of no culinary interest.

August–October

Suillus luteus

Dark brown slimy cap with tiny yellow pores on underside, and a white ring on stalk.

The stalk is often covered with brownish granules above the ring, becoming white or brown below. The thick flesh is pale yellow and does not change colour when bruised. Good to eat if the slime is first removed.

Common and widespread, especially under pines.

Suillus spp have slimy caps. *S. granulatus*: pale brown cap, pores exude fluid, no ring, edible after peeling; *S. bovinus*: yellowish cap, large pores, no ring, **inedible**; *S. grevillei*: under larch, golden-yellow cap, ring, **inedible**.

July–November

Mycena epipterygia

Delicate pale brown mushroom with a bell-shaped slimy cap, striated at edge, and a slender, yellowish slimy stalk. Spore print white.

Gills white with a yellow tinge. **Nitrous Mycena** (*M. leptocephala*) (**1**), growing in clusters on pine needles, is not slimy, cap opens out more and is greyer. Stalk grey. Gills pale, spores white.

Both are very common and widespread under conifers.

Milk-drop Mycena and many other small brown mycenas. *M. sanguinolenta*: stem exudes a red juice. Mycenas have fibrous stems, attached gills, often sinuate, and white spores. Cap colour ranges from brown to white, pink, lilac or yellow.

September–November

Russula xerampelina

Brittle flesh and gills. Cap pinkish purple, blackish in centre. Gills pale ochre, becoming russet. Flesh turns yellow when cut and smells of crab.

The very variably coloured cap pales towards the edge, becoming pink or carmine. Cap becomes sunken in centre when mature. Stalk tinged with red, yellowing towards the base. Spore print cream to ochre. Edible.

Widespread and common under pines, also found under spruce.

Similarly coloured caps, under conifers: **Purple Russula**; **R. atrorubens**; *R. turci*: cap with distinct ring of colour around darker centre, under pine and spruce. *R. graveolens*: smells of crab, under broadleaved trees.

August–November

R. drimeia (**1**), cap 10–15cm:4–6in, bright purple-violet to wine red, gills bright lemon becoming creamy. Under pine on sandy soils. *R. queletii*, especially under spruce in damp places in mountains, is very similar with a distinctive reddish stalk. **Both inedible**.

Purple Russula *R. amara* (*caerulea*) (**2**) cap 8–10cm:3¼–4in, with a conical outline when young that is unusual for a russula, dark violet to slaty purple. When fully expanded always has an umbo in centre. Stalk white, gills yellowish. Spore print yellowish. Skin of cap very bitter, making it **inedible**.

R. atrorubens (**3**), cap 6–8cm:2½–3¼in, pinkish red at margin, becoming darker purplish in centre. Gills and spore print white. Stalk white. Faint fruity smell. Flesh very acrid. Under conifers, especially in damp places.

Lactarius deliciosus

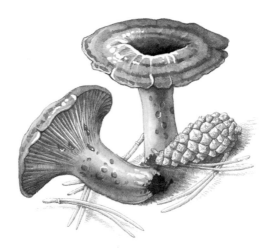

Cap with concentric orange rings on a paler ground, turning green on bruising. Orange gills. Flesh and gills release a sweet orange milk with a slightly peppery aftertaste.

A large thickset fungus, becoming vase-shaped with age. Cap has slightly frosted or shining surface and a thin inrolled edge. Stalk pale, thick, with orange pits. Flesh off-white. Gills crowded. Spore print white. Edible.

Widespread and fairly common under pines on neutral or limy soils.

Red-milk Milk-cap; *L. deterrimus*: cap uniformly pale orange, stalk unpitted, under spruce only; *L. salmonicolor*: orange markings on cap less regular, slightly soapy smell, bitter taste but edible, under firs (*Abies*) only.

August–October

41

RED-MILK MILK-CAP
Cap 5–10cm:2–4in; stalk to 5cm:2in

Lactarius sanguifluus

Pinkish orange gills release orange milk which immediately turns blood red. Cap dull orange, frosted, with pink and orange zones. Stipe pitted reddish orange.

Edible, with a mild taste. Flesh white, immediately reddening when cut. Spore print white. **Red-hot Milk-cap** (*L. rufus*) (**1**) cap 5–10cm:2–4in, has a brick red cap, abundant white milk, cream to yellow arched gills and a very acrid taste.

Red-milk Milk-cap: (not found in Britain) under pine in southern Europe; Red-hot Milk-cap: common under conifers, also under birch.

Saffron Milk-cap; *L. deterrimus*: milk turns red only slowly; *L. semisanguifluus*: flesh reddens on cutting, then turns green in 24 hours, cap often with a green tinge, stalk not pitted, edible.

August–November

42

Lactarius camphoratus

Dark chestnut brown cap, fading to reddish. Strong smell of "wet linen" when fresh. Flesh and gills release a watery milk. Gills pinkish yellow. Spore print cream.

A small slender lactarius. When dried the flesh has a strong spicy smell resembling curry powder and can be used as a seasoning. The cap often becomes funnel-shaped but retains a small central umbo.

Quite common and widespread, growing in clusters often at the base of trees, mainly with conifers but also in association with broadleaved trees.

Red-hot Milk-cap: generally larger, acrid milk.

August—November

SAFFRON PARASOL
Cap 2–5cm:¾–2in; stalk to 7cm:2¾in

Cystoderma amianthinum

Yellow cap has a granular surface and remnants of veil around edge. Stalk is covered in granules up to the ring, which points upwards, and is then smooth above. Spore print white.

Gills adnate, white to cream. The whitish flesh has a "mouldy" smell. **Pine Spike-cap** (*Chroogomphus rutilus*) (**1**) has a sticky red-brown cap with a central "spike", a small ring on the stalk, purplish grey gills and brown spores.

Saffron Parasol: very common and widespread under conifers, in mixed woods, in grass, and on heaths and moors. Pine Spike-cap: common, especially under pines.

C. carcharias: paler and greyer in colour, rare, mainly in mountains; *C. granulosum*: red-brown cap and stalk; *Lepiota clypeolaria* and other similar small lepiotas: scaly cap, stalk shaggy below ring, gills free, spore print white.

August–October; August–November (**1**)

FALSE MOREL

Cap 5–10cm:2–4in; stalk to 4cm:1½in

Gyromitra esculenta

1

Rusty brown convoluted cap like a brain, set on a short stout paler stalk.

The pale brittle flesh has an aromatic smell. **Deadly when raw** and should be avoided. **Cauliflower Mushroom** (*Sparassis crispa*) (**1**) has pale beige or creamy flat, curled, ribbon-like divisions. Edible.

False Morel: under pines, especially in hills, more common in northern Europe. Cauliflower Mushroom: at foot of pines, sometimes on stumps, quite common.

False Morel: *G. infula*, cap less lobed, stalk longer; **Morel**. Cauliflower Mushroom: *Ramaria* and *Clavulina* spp. have coral-like branched fruitbodies and are **inedible**, although not poisonous; **Bear's Head Tooth**.

April; August–November (**1**)

Cantharellus cibarius

Golden-yellow to apricot throughout smelling faintly of apricots. The top-shaped cap soon becomes funnel shaped, and has thick forking ridges on underside instead of true gills.

A delicious edible fungus, often found growing in troops. **Funnel Chanterelle** (*C. tubaeformis*) (**1**) has a yellow stalk and forking yellow or greyish ridges on underside of cap, which is grey-brown or entirely yellow on top. Good to eat.

Both the Chanterelle and Funnel Chanterelle are quite common in woods, especially oak and beech, and also under conifers.

C. lutescens: golden-yellow ridges and stalk, greyish top, delicious; **False Chanterelle**: gills, harmless; **Jack O'Lantern**: on wood, gills, **poisonous**; *Gomphus clavatus*: inverted cone-shaped, ridged outer surface violet.

July–December

Hydnum repandum

Fleshy lobed cap pale fawn or orange-tinged, dark russet in one form, with cream to russet teeth on underside instead of gills. Stalk white, with faint bloom, often misshapen.

A good edible fungus with a nutty taste after cooking to remove bitterness. Teeth are brittle, crowded and run down the stalk a little way. The flesh smells faintly of orange-flower water.

Quite common and widespread under broadleaved trees and conifers.

Sarcodon imbricatum: scaly cap, under conifers; *Bankera fuligineoalba*: smaller, less widespread, pinkish tinged cap, under conifers; *Phellodon niger*: dark blue-grey cap, under conifers; *P. tomentosus*: russet brown cap, paler at edge.

August–November

Boletus edulis

Date-brown fleshy cap with a greasy surface and white edge.
Pores pale, slowly becoming greenish yellow. Upper stalk covered
with faint network of white.

One of the best-known edible fungi, the most delicious of the
boletes, retaining its flavour on drying. The stalk is pale, streaked
reddish brown in places. The pores do not change colour on
bruising.

Quite common and widespread under broadleaved trees, in mixed
woods, glades or copses.

Spring Bolete; **Bay Bolete** and other boletes. Avoid the **Bitter
Bolete**. Also avoid boletes with red pores, or dark brown lobed
cap (*B. fragrans*), or with flesh turning blue, some of which are
toxic.

August–November

Xerocomus (Boletus) badius

Cap rich bright brown, pores cream to lemon, bruising greenish when mature, stalk cylindrical rather than swollen at base as in some other boletes.

Stalk pale brown streaked with reddish brown. Pale flesh of cap blues slightly to the touch. Edible and good. **Red-capped Bolete** (*Krombholziella aurantiaca*) (**1**) is large and fleshy, with orange brown cap, pale pores, scaly stalk. Edible.

Bay Bolete: common and widespread under conifers and broadleaved trees. Golden Bolete: rarer, and found under poplar and birch.

Red-cracked Bolete and lookalikes; **Brown Birch Bolete**; avoid boletes with pale caps, or red pores or those whose flesh turns blue, some of which are **toxic**.

August–November

Gyroporus (Boletus) cyanescens

White to pale ochre cap is smooth and velvety becoming scaly.
Pores white to pale ochre, bluing when bruised. Flesh turns
instantly blue when broken.

An edible bolete despite the bluing flesh. Stalk is white at top,
brownish from the base up to an indistinct ring-like zone which
soon disappears, and becomes hollow at maturity.

Quite widespread but not very common, under broadleaved trees
and conifers, especially birch and spruce on sandy soil.

Chestnut Bolete: chestnut cap, yellowish pores, flesh unchanging,
lumpy brown stalk, **possibly toxic**. *Suillus laricini*: under larch,
olive grey cap, ring on stalk, flesh greens slightly when cut.

July–November

Tylopilus (Boletus) felleus

Large bolete, with dry grey brown cap and thick stalk, covered in a darker network. White pores becoming salmon pink when older.

The very bitter flesh makes this bolete **inedible**, although not toxic, and can spoil a dish of ceps if even one is included in error.

In both broadleaved and coniferous woods, common locally but not widespread throughout Europe.

Cep; **Spring Bolete**; **Bay Bolete** and other **boletes**.

June–November

Strobilomyces strobilaceus (floccopus)

Cap covered with thick grey brown to black scales, resembling an immature pinecone when young. Stalk with a shaggy ring, becoming indistinct in older specimens. Spores purple-brown.

This unusual bolete has whitish pores covered with a veil when young, later becoming greyish. Flesh reddens on bruising. **Inedible**.

Uncommon, often found under beech, and also with conifers.

Unlikely to be confused with any other bolete. Its purple spores are an unambiguous distinguishing factor, other boletes having whitish spores.

September–November

Collybia dryophila

1

Smooth bright red-brown cap becoming yellow with age. Crowded white gills, adnexed. Bright orange-brown stalk, paler at top, smooth. Spore print white.

Slightly poisonous when raw, so do not eat. **Wood Woolly-foot** (*C. peronata*) (**1**) has a leathery yellowish brown cap and stalk, which is woolly at the foot. Bruised flesh smells of vinegar. White spores. Not toxic but not edible.

Common and widespread, Russet Shank especially under oak and beech, and on heaths, amongst bracken, Wood Woolly-foot in woods of all kinds.

Collybias have yellowish, white or brown caps, tough stalks, gills adnate to free, spores white or pinkish. *C. confluens*: pinkish downy stalk; **Spindle Shank**: in clusters under beech or oak, brown cap, greyish gills, swollen stalk tapers at base.

July–November; September–November (**1**)

Mycena pura

Small fleshy rose-pink to lilac mushroom with white to pinkish gills. Smells of radishes. Fibrous stalk.

One of the more robust mycenas, the bell-shaped cap at first becoming convex to flat, sometimes with a central knob. Edge is striated. Gills adnate, spaced quite far apart with short gills in between at cap edge. Spore print white.

Common and widespread, occurring in woods of all kinds, especially beech and oak, but also under various conifers.

M. pelianthina: dark edge to gill; **Milk-drop Mycena**; mycenas have fragile bell-shaped caps in various colours, tough stalks, adnexed/adnate to sinuate gills, white or pink spore prints; **Amethyst Deceiver**: thick violet gills.

June–December

Mycena galopus

Small greyish-brown fragile mushroom, with a hemispherical cap striated at edge. Stalk exudes white milky fluid when broken. Gills white, spore print white.

Cap colour can vary from whitish to almost black. Gills adnexed. The slender fibrous stalk is smooth with minute hairs at the base, resembling a root.

Common and widespread, found growing amongst leaf litter in all kinds of woods, never on living trees.

One of numerous small brown mycenas, this can be distinguished from most of them by its white milk. *M. metata*: pale pinkish brown cap, no milk, smells of iodoform; *M. filopes*: no milk, smell of iodoform, usually in moss.

July–December

COMMON WHITE INOCYBE
Cap 1–3cm:½–1¼in; stalk to 5cm:2in

Inocybe geophylla

Cap and stalk white and silky, stalk powdery at apex only. Gills crowded, yellowish to clay brown, adnexed. Spore print brown.

One of a number of **very poisonous** small white mushrooms. Cap colour can vary from white to lilac with a pale ochre tinge in centre.

Common and widespread in all kinds of woods, often growing in damp soil by paths.

Inocybes often have conical caps (e.g. **Deadly Inocybe**), or flatter caps with a central knob, often radially streaked and splitting at edge. Most are some shade of brown. Gills and spores brown. **All should be avoided, many are toxic.**

June–November

Paxillus involutus

Cap olive brown to rusty brown, felty, sticky when wet, with inrolled rim. Gills easily separable from cap, cream to dirty yellow-brown, crowded, running down stem. Spores dull brown.

A robust fungus which could at first sight be confused with some lactarias but is not milky. Gills spotted rust brown. Stalk pale brown, thick and smooth. **Not to be eaten** as it can have a cumulative toxic effect.

Very common in mixed woods, often found under birch and oak.

P. atrotomentosus: velvety black stalk, on conifer stumps; *Gomphidius glutinosus*: sticky grey-brown cap, whitish stalk with a constriction and ring-like zone immediately beneath gills, under conifers.

August–November

57

THE SICKENER
Cap 3–10cm:1¼–4in; stalk to 7cm:2¾in

Russula emetica

Brittle gills and flesh, pure scarlet or cherry-red shiny cap, white gills and stalk. Skin of cap peels completely to show white or faintly reddish flesh. Very acrid taste.

One of numerous bright red-capped russulas, this can cause vomiting in the raw state and **should be avoided**. The flesh has a distinctive smell of coconut. Gills adnate, quite widely spaced. Spore print white.

Common and widespread under both conifers and broadleaved trees.

The red-capped, white-spored russulas are difficult to distinguish from each other and are **all best avoided**. Some red-capped russulas have cream gills, ochre to yellow spore print and a mild taste. None is worth eating.

July–October

Russula nigricans

Brittle gills and flesh. Whitish to grey-brown cap and stalk turn black with age. Flesh reddens (blackens in old caps) on cutting. Gills thick, whitish, with short gills in between.

The large blackened caps persist for a long time. Spores white. Edible but not worth eating. **Foetid Russula** (*R. foetens*) (1), cap 12–20cm:4¾–8in, has a dirty yellow-grey slimy cap and a sickening smell. Gills whitish, spores pale cream.

Both are common in mixed woods, especially broadleaved woods.

Blackening Russula: *R. albonigra*, rarer, gills crowded, flesh blackens without reddening; *R. delica*, remains white to pale brownish; *R. densifolia*, crowded gills. Foetid Russula: *R. laurocerasi*, rarer, smells of marzipan.

June/July–October/November

Russula ochroleuca

Golden-yellow cap, becoming sunken in centre. Gills white, stalk greyish. Taste variable, from peppery to mild. Spore print pale cream.

Cap often tinged with ochre, orange or green. **Grey-blue Russula** (*Russula parazurea*) (1) has a distinctive matt blue-grey cap with a slight bloom towards the edge. Gills and spore print cream. Edible.

Common Yellow Russula: very common and widespread; Grey-blue Russula: common, typically found in oak and beech woods.

Common Yellow Russula: *R. fellea*, straw-coloured, beech and oak woods; *R. claroflava*: brighter yellow, yellow gills, under birch; Grey-blue Russula: *R. ionochlora*, grey cap with greenish or violet tints, under beech.

August–November

DIRTY MILK-CAP

Cap 5–12cm:2–4¾in; stalk to 6cm:2½in

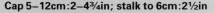

Lactarius plumbeus (turpis)

Flesh and gills release white acrid milk when broken. Slightly
slimy olive brown to blackish cap, rolled under at edge. Yellowish
gills, bruising brown. Stem grey-brown.

Large squat fleshy mushroom, cap edge slightly velvety when
young. Gills crowded, running down stem a short way, soon
becoming spotted with grey. Stalk stout and tapering to base.
Spore print cream. **Inedible.**

Common and widespread in damp soil under birch, also under
pine.

Slimy Milk-cap: concentric rings of spots on cap, under beech; *L.
fluens*: white stalk, under beech; **Brown Roll-rim**: not milky.

July–November

AMETHYST DECEIVER
Cap 1–8cm: ½–3¼in; stalk to 10cm:4in

Laccaria amethystea

Vivid violet throughout when young, fading bluish or whitish with age. Gills thick, far apart, adnate to decurrent. Spore print white.

The cap has a matt surface covered with fine scales and soon flattens out. Stem fibrous, usually curved. Harmless but not worth eating.

Very common everywhere in shady woods and damp places.

Lilac Mycena; Deceiver.

August–November

SPOTTED COLLYBIA

Cap 5–10cm:2–4in; stalk to 12cm:4¾in

Collybia maculata

1

Smooth white cap with rusty pink spots. Crowded white gills, adnate or nearly free. Fibrous white stalk, sometimes with reddish stripes. Pale pinkish buff spore print.

Bitter and indigestible, should not be eaten. **Broad-gilled Agaric** (*Megacollybia platyphylla*) (**1**) cap 6–15cm:2½–6in, is a translucent grey-brown, streaked with dark fibrils. The white stalk ends in "roots". Spores white. Not worth eating.

Common and widespread, Spotted Collybia mainly under birch or conifers, and on heaths and moors, Broad-gilled Agaric in leaf litter and decaying wood of beech, oak, more rarely conifers.

Cloudy Agaric; Sweating Mushroom: smaller, decurrent gills, white spore print, **poisonous**; other collybias have reddish, brown, white or yellowish caps, tough stalks and adnate to free gills, spore print white or pinkish.

July–November; September–November (**1**)

63

CLOUDY AGARIC
Cap 12–20cm:4¾–8in; stalk to 10cm:4in

Clitocybe nebularis

1

Fleshy cap convex, becoming flatter with central hump.
Yellowish grey cap and stalk, gills thin, crowded, cream to
yellowish, slightly decurrent. Spore print white.

Do not eat as it disagrees with some people and can also be
confused with poisonous species. **Club-footed Clitocybe** (*C.
clavipes*) (**1**) has a club-shaped downy base to stalk, and watery
flesh smelling of bitter almonds. Not worth eating.

Common and widespread in both coniferous woods and under
broadleaved trees.

Other larger clitocybes: *C. geotropa*, smells of new-mown hay; *C.
inornata*: pale coffee coloured cap, smell becomes unpleasantly
fishy. Clitocybes are fleshy, white-spored, with decurrent to
adnate gills and white, brown or greyish caps.

August/September–December

SHORT RHODOCYBE
Cap 4–12cm:1½–4¾in; stalk to 8cm:3¼in

Rhodocybe truncata

Fleshy pale pinkish brown to russet cap with irregular inrolled edge. Gills decurrent, cream to similar colour as cap. Spore print pink.

The cap has a felty surface. Gills are crowded, forked and narrow. The stem is thick and solid, whitish tinted pink. Edible.

In mixed woods and grassy thickets, often growing in clusters or rings.

Poison Pie: toxic, and other *Hebeloma* spp, **none of which should be eaten,** brownish spore print; èntolomas, some of which are **deadly,** also have pink spore prints; **The Miller.**

September–November

WOOD BLEWIT
Cap 10–15cm:4–6in; stalk to 10cm:4in

Lepista (Clitocybe) nuda

Distinctive large smooth bluish lilac cap when young, becoming tan. Gills sinuate, lilac becoming yellowish, and can be separated easily from cap. Spore print pinkish buff.

This common mushroom is good to eat after cooking (although some people are allergic to it) but is **slightly poisonous when raw**. Stalk stout, fibrous, lilac. The flesh is thick, firm and white with a lilac tinge.

Common and widespread in open woods, along paths, in gardens, often growing in groups.

Blewit (*L. saeva*): buff cap, pinkish beige gills, lilac stalk, in grass and wood edges, edible. Make sure you do not have **Lead Poisoner** or a **clitocybe**, which also have pinkish spore prints, or a **tricholoma** (white spores).

September–December

Hebeloma crustuliniforme

Sticky cream to russet cap with paler margin. Gills crowded, pale brownish grey to cinammon brown, edged with water droplets when young. Spore print brown to rust.

Poisonous. This mushroom smells of radishes when young, has a bitter taste and **can cause severe gastric upsets**. The stout cylindrical stalk is white, flaky, with white granules on surface towards apex.

Quite common and widespread in deciduous and mixed woods and in grass, sometimes forming fairy rings.

Hebelomas have fine-toothed gill edges (visible under a hand lens) and brown spores, and usually sticky caps in shades of brown. **All are poisonous**. **Tricholomas**: white spores; **entolomas**: pink spores.

August–November

67

Marvellous Tricholoma
(*Tricholoma portentosum*) (**1**) cap
10–15cm:4–6in, grey-brown,
slightly conical, tinged
yellowish, with dark fibrils
radiating out from blackish
centre. Sinuate gills and stalk
white, tinged lemon. Spore print
white. An excellent edible
mushroom with a floury smell.

White Tricholoma (*T.
columbetta*) (**2**) has a satiny white
surface to cap, sticky when
damp, and white sinuate gills
and stalk, cap and stalk
discolouring with blue or pink
spots. White spores. Under
beech or birch. Edible.

Leopard Tricholoma (*T.
pardinum*) (**3**) cap 15–25cm:6–
10in, large mushroom with a
fleshy cap incurved at edges,
covered with brownish scales on
a paler ground. Stout smooth
pale stem. Gills whitish. Earthy
smell becoming unpleasant.
Spore print white. **Highly
poisonous**. Not found in
Britain. Mainly in beech and fir
woods.

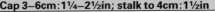
Hygrophoropsis aurantiaca

Bright golden yellow to orange fungus, cap soon sunken in centre. Gills forked, bright orange, running down stalk. No distinctive smell. Spore print white to cream.

This harmless Chanterelle lookalike is edible but not really worth eating. The surface of the cap is matt, dry and sometimes downy, and eventually becomes brownish. The flesh is yellowish and rather tough.

Common and widespread, often occurring in large troops, under both coniferous and broadleaved trees.

Chanterelle; Jack O'Lantern: poisonous, grows on wood.

August–November

69

Cortinarius purpurascens

Brown ring-like remnants of veil on stalk. Slimy cap is reddish brown to greeny-greyish, violet at edges. Gills lilac, becoming rusty with age. Spore print rust brown.

Large fleshy mushroom in which edge of cap is joined to stalk by a fine veil when young. The stout pale violet stem has a bulbous base with a distinct rim. Harmless but **inedible**. Take care not to mistake it for the edible Wood Blewit.

Widespread but not abundant, in clumps, in coniferous and mixed woods.

C. alboviolaceus: pale silky cap, lilac gills, under beech, oak; *C. violaceus* (**1**): (not found in Britain) deep violet cap, gills and stalk; *C. speciosissimus*: conical brown cap, brown stem and gills, **deadly**, coniferous woods, moors.

September–November

RED-GILLED CORTINARIUS

Cap 5–8cm:2–3¼in; stalk to 7cm:2¾in

Cortinarius semisanguineus

Reddish-brown ring-like remnants of veil on stalk. Cap velvety, pale yellowish brown to olive brown, with a central knob. Blood red crowded adnate gills. Spore print brown.

One of several cortinarias with red gills. Edge of cap attached to stalk by a fine veil when young. Stalk yellowish. Flesh pale, yellowish, smelling faintly of radishes. Not edible and **may be toxic**.

Quite common and widespread in coniferous and birch woods, and on heaths, often growing in clusters.

C. sanguineus: blood red cap, gills and flesh, common; *C. cinnabarinus*: similar, more orange-red, rarer, under beech, hornbeam; *C. orellanus*: **deadly**, orange to rusty brown gills, orange-brown cap, rare, in warm places.

August–November

Stropharia aeruginosa

Slimy cap is a bright bluish green when young, becoming yellowish with age, and has white fleecy scales at the edge. The stalk has a greyish ring. Gills adnate, violet-grey.

The slimy stalk is smooth above the ring, fleecy or covered with white flecks below, and paler than the cap. The vivid colour of the cap soon fades, and the gills become chocolate brown. Spore print purplish-brown. **Inedible**.

Quite common and widespread in grassy woods and copses, gardens and pastures.

Other stropharias have beige or brown caps, whitish stalks with a ring, purple-brown spores, greyish violet adnate or slightly decurrent gills. The very common *S. semiglobata* with a yellowish hemispherical cap grows on dung.

June–November

THE GYPSY

Cap 8–12cm:3¼–4¾in; stalk to 10cm:4in

Rozites caperata

Pale golden to dull ochre-yellow cap sometimes with silvery veil covering centre. Gills pale becoming clay-brown, edges finely toothed (under a hand lens). Whitish ring on stalk.

Pale brown spores. Stalk pale, striated, ring fleshy and remaining on stalk for a long time. Gills adnate, thick and crowded. Very good to eat if mushrooms undamaged by insects can be found.

Fairly common under birch and pine in mountainous woods on acid soil in northern Europe.

Poison Pie and other *Hebeloma* spp: no ring or very indistinct ring-like remains of veil on stalk.

August–December

THE PRINCE
Cap 10–25cm:4–10in; stalk to 20cm:8in

Agaricus augustus

A large agaric, the cap covered with small golden to dark brown scales. Stalk bruises yellow. Ring large and hanging. Flesh discolours brownish, smells of bitter almonds.

A good edible mushroom. The young cap is hemispherical, expanding flat with age. Stalk is scaly below ring. Greyish to dark brown gills very crowded. Spore print dark brown.

Quite common under various broadleaved trees and conifers.

Yellow-staining Mushroom: white silky cap, stains yellow at base of stalk, **toxic**; *A. silvicola*: white cap, stains yellow where bruised, in mixed or broadleaved woods, edible; most other mushrooms are found in open grassland.

July–October

FLY AGARIC

Cap 10–20cm:4–8in; stalk to 15cm:6in

Amanita muscaria

Unmistakable bright red cap covered with small white patches of the universal veil. White gills and spores. White stalk with ring. Volva only a series of ridges at base of stalk.

A familiar mushroom in northern Europe, illustrated in countless fairy-tales. Young fruitbodies are entirely covered in a white veil. The white patches may become washed off in older specimens and caps fade orange-red. **Poisonous.**

Very common under birch or pine on poor soils.

The edible and good **Caesar's Mushroom** (not found in Britain) has a smooth red cap with no or larger remnant of white veil, yellow stalk and gills, and a large sac-like white volva, and has a more southerly distribution.

August–November

THE BLUSHER
Cap 10–18cm:4–7¼in; stalk to 10cm:4in

Amanita rubescens

Light brown cap covered with small dingy yellowish patches of veil. Gills and spores white. Stalk white flushed pink with hanging ring. No apparent volva. Flesh white, flushing pink.

This is one of the few edible and excellent amanitas, but must be cooked as it is **indigestible when raw**. Gills are speckled pink in older specimens. **Do not confuse with The Panther**.

Common and widespread under all kinds of trees.

The Panther: deadly poisonous, white scales on cap, basal bulb with remains of volva, flesh does not turn pink; *A. spissa*: dark brown cap with greyish scales, flesh does not turn pink, not poisonous but not worth eating.

July–November

Amanita citrina

Pale lemon yellow cap, sometimes almost white, with loose patches of the white veil adhering to it. Margin not striated. Bulbous base with volva. White ring on stalk.

This amanita is not poisonous but **should not be collected for eating** for fear of confusion with the Death Cap. It smells of raw potatoes and has an unpleasant taste. Gills and spore print white.

Widespread and quite common, under broadleaved trees and conifers. Also found on heaths.

Other amanitas, especially **Death Cap** and **Jonquil Amanita**.

August–October

FLUTED BLACK HELVELLA
Cap 3–5cm:1¼–2in; stalk to 6cm:2½in

Helvella lacunosa

Dark grey thin saddle-shaped cap, with wavy lobes, and with no gills, pores or ridges on underside. The hollow paler grey stalk is deeply fluted and furrowed.

A distinctive fungus with upper surface of cap sometimes slightly wrinkled, lower surface slightly paler and smooth. Eaten in some countries but **not recommended** as related fungi are known to contain toxins.

Quite common and widespread in coniferous and broadleaved woods, especially on burnt soil.

Leptopodia atra: smaller, with a round lobed cap and slender blackish stalk; **Fluted White Helvella**: whitish to beige cap, fluted white stalk; **Common Morel**: rounded cap with honeycombed surface.

September–October

Phallus impudicus

Unmistakable, recognizable by its shape and strong offensive smell. The stout spongy white stalk carries a slimy foetid green-black mass of spores at its top.

Often first detected by the unpleasant smell. The slime and spores are eventually eaten by flies, exposing the pitted surface of the cap. The immature fruitbody is like an "egg", its remnants forming a volva-like sac at stalk base.

Common and widespread in woods and gardens, especially on rich soil.

P. hadriani: similar, much rarer, on dunes by the sea, spore mass has faint sweetish smell when still in the "egg"; *Mutinus caninus*: similar shape and structure, much smaller, stalk yellowish, spores borne direct on tip of stalk.

July–November

COMMON EARTHBALL
Fruitbody 6–12cm:2½–4¾in diameter

Scleroderma citrinum

Hard ball-like fungus growing directly on the ground, with a dirty yellowish scaly outer covering, becoming cracked, and a powdery purplish spore mass inside.

This very common earthball has no stalk and "roots" directly into the ground. The internal mass of spores is initially pinkish, eventually turns purple and escapes through cracks in the thick hard outer wall.

Very common in woods, parks and heaths, especially on mossy peaty ground and under birches.

S. bovista: thinner, smooth outer wall, short stalk-like base very earthy, less common; *S. areolatum* and *S. verrucosum*: surface more finely scaled, base has a short tapering stalk, spore mass olive-brown, common.

August–December

80

EARTHSTAR

Fruitbody 8–12cm:3¼–4¾in diameter

Geastrum triplex

Yellowish grey-brown sac containing spores sits in a saucer supported on curved-back segments of fruitbody outer wall, which has split radially.

In this earthstar, the rays are rather fleshy and cracked across. It also has a distinct halo around the small opening in the inner sac through which the spores are released.

Quite common and widespread in woods.

G. sessile: rays not cracked across, no saucer at base of inner sac; *G. fornicatum*: rays form a four-legged support 5–10cm:2–4in high, rare; *G. quadrifidum*: similar to latter but smaller; *Astraeus hygrometricus*: rays spread out.

August–October

81

CRESTED CORAL
Fruitbody 3–10cm:1¼–4in across

Clavulina cristata

White coral-like fruitbody, branching, with expanded toothed tips to branches. Flesh white.

One of a type of fungus commonly called Fairy Clubs or Coral Fungi. Some related species have thick unbranched club-like fruitbodies. The Crested Coral is edible, but some other species are not.

Common on the ground in woods of all kinds, related species also occur in grass or grow directly on wood.

C. cinerea: ash-grey but otherwise very similar; *Ramaria* spp: generally more massive, lilac, yellow, pale pink or olive, not pure white; *Clavulinopsis corniculata*: yellow, tips of branches pointed, not toothed.

June–July

82

Small colourful (yellow, red, orange, pink) mushrooms which grow in grass, appearing in late summer and autumn. All have rather thick "waxy" gills and white spores.

Parrot Mushroom (*Hygrocybe psittacina*) (**1**): cap 2–5cm:¾–2in, sticky, shades of brownish yellow tinged greenish and/or reddish. Stalk tough, slimy, green at apex. Gills adnate, yellowish. Heaths, pasture, grassy woods. *H. ceracea* is waxy yellow, cap sticky, gills decurrent. *H. punicea* is much larger, cap 5–12cm:2–4¾in, blunt-topped blood-red, sticky when damp. Gills adnexed, yellow tinted red.

Orange Waxy Cap (*H. miniata*) (**2**): cap 1–2cm:½–¾in, scarlet or orange-red, not slimy, with small scales in centre. Gills adnate-decurrent, orange with yellow edges. Likes damp places. *H. coccinea* has a blood-red bell-shaped cap, yellow gills.

Conical Slimy Cap (*H. conica*) (**3**): cap 1–4cm:½–1½in, orange-red, blackening. Gills yellowish, free. Heaths, pasture, grassy woods. *H. nigrescens*, also blackening, larger, white base to stalk.

Psilocybe semilanceata

1

Narrow pointed pale yellow-brown cap with a greasy surface, never expanding, and pale wavy stalk. Gills adnexed, brown with a white edge. Spore print purple-brown.

Like some other psilocybes this contains the hallucinogenic drug psilocybin and **can cause delirium**. **Mountain Moss Psilocybe** (*Ps. montana*) (**1**) is dark brown, with a hemispherical cap, opening out flat with a central umbo.

Liberty Cap: common in grassland, parks, by the roadside. Mountain Moss Psilocybe: common on heaths and moorland, but also grows on rich soil by the roadside.

Avoid all "little brown mushrooms"; **some are deadly**. **Psathyrella**: black spores; **Lawn Mower's Mushroom**; *Bolbitius*: shrivels rapidly, brown spores. **Mycena**: white spores; **Deadly Galerina**: brown spores.

August–November; April–October (**1**)

Clitocybe dealbata

Small creamy white mushroom, with pinky brown markings on older caps. Gills thin, adnate-decurrent, creamy, crowded. Stalk white and silky. Spore print white.

A **highly poisonous** species which could be gathered in mistake for several edible mushrooms. It contains the toxin muscarine which produces sweating, blurred vision and involuntary muscle movements.

Common on lawns and in grass generally.

There are several white clitocybes, characterized by thin decurrent gills and white spores, e.g. *Clit. rivulosa*: wrinkled cap, **toxic**. **The Miller**: pink spores; **Buff Meadow Cap**: thick gills, white spores.

August–November

FAIRY-RING CHAMPIGNON
Cap 2–5cm:¾–2in; stalk to 7cm:2¾in

Marasmius oreades

A common mushroom forming large rings in lawns and pastures. Moist, bell-shaped rusty brown cap flattens out and pales to yellowish brown. Stalk tough, felty. Spore print white.

Like all *Marasmius* species, this tough mushroom quickly revives on moistening after drying out. Good to eat if the tough stalks are discarded but **take care not to confuse** with poisonous whitish clitocybes also common in grass.

Very common in lawns, pastures and by paths.

Sweating Mushroom & *Clit. rivulosa*; **Mycena** species also have white spores.

May–November

Cap 1–6cm: ½–2½in; stalk to 8cm: 3¼in

Laccaria laccata

Reddish brown to pinkish brown smooth cap soon becoming sunken in centre. Gills pale with a pinkish tinge, thick, slightly waxy and spaced quite far apart. Spore print white.

A variable mushroom, common but often hard to identify. Gills adnate. **Inedible. Lawn Mower's Mushroom** (*Panaeolina foenisecii*) (**1**) has a brown bell-shaped cap, 1–2cm: ½–¾in, drying paler in centre. Brown gills adnate, mottled. Spores brown.

Deceiver: very common and widespread on heaths and in broadleaved woods, also in sphagnum bogs. Lawn Mower's Mushroom: very common in lawns and short grass.

Amethyst Deceiver: violet throughout when young and moist, drying paler, thick gills spaced far apart and white spores, in woods; **Mycena** spp; **Psathyrella** spp; *Hygrophorus* & **Hygrocybe** spp. have waxy gills and white spores.

July–December

BUFF MEADOW CAP
Cap 5—10cm:2—4in; stalk to 5cm:2in

Cuphophyllus (Camarophyllus) pratensis

A rather stout squat mushroom, with fleshy buff to dull orange cap. The pale buff gills are waxy, widely spaced and run down and merge with stalk. Spore print white.

An excellent edible mushroom. The surface of the cap is dry and matt and often cracks around the centre. The thick gills are thin-edged and have short gills interspersed between them. The stalk is stout, and a similar colour to gills.

Common and quite widespread in meadows, lawns, grassy copses and also in woods.

Chanterelle: ridges on underside of cap, not gills, orange-yellow; **False Chanterelle**; *Cuph. niveus* and others: white, smaller, edible but **best avoided** for fear of confusion with toxic white clitocybes (e.g. **Sweating Mushroom**).

August—December

ST GEORGE'S MUSHROOM
Cap 5–13cm:2–5¼in; stalk to 8cm:3¼in

Calocybe gambosa

White to greyish buff fleshy convex cap with an inrolled edge.
Gills white to cream, narrow and crowded, sinuate. A
"mushroom" smell. Spore print white.

This mushroom is only found in the spring, never in the autumn.
Flesh firm and white. An excellent edible mushroom.

Quite common and widespread in grassy woodland glades, wood
edges, under hedges. Sometimes forms large fairy-rings in
grassland.

Take care not to confuse with toxic white **clitocybes, Lead
Poisoner** or **Deadly Inocybe**, which grow in similar situations,
but which usually appear later.

April–June

FRIED-CHICKEN MUSHROOM
Cap 4–12cm:1½–4¾in; stalk to 12cm:4¾in

Lyophyllum decastes

Growing in clusters on ground often around stumps or buried roots. Smooth shiny yellowish grey to reddish brown cap, crowded white adnate gills. Stalk whitish. Spore print white.

Although sometimes considered edible, this is **best avoided**. The caps begin convex and then flatten out and the gills become straw-coloured with age. The thin white flesh is tough with a faint mealy smell.

Common and widespread in grassy copses, gardens, parks.

L. fumosum: greyish gills, stems joined together at base, on rich soil, compost heaps; *L. connatum*: glossy white caps, white arched decurrent gills, white stalk, in clusters in grass; **Lead Poisoner**.

July–October

Psathyrella lacrymabunda

Brown to blackish gills "weep" black droplets. Grey-brown velvety or scaly cap with a woolly edge often blackened by spores. Ring-like marking from veil on stalk. Spores black.

Gills adnate, mottled, with white edges. Fibrous stalk. Edible. One of the larger psathyrellas, some others are very small and delicate. The edge of young cap is attached to stalk by a fine veil in some, but not all, psathyrellas.

Very common and widespread in fields, on bare soil, beside paths, often in association with buried dead wood or roots.

Ps. candolleana: creamy white cap, splitting at edges, gills pale brownish lilac, stalk white, no ring, common on and around stumps, in nettle-beds, copses; **Clustered Psathyrella**; **Cortinarius** species: brown spores; **Deceiver**.

April–November

Agrocybe praecox

Beige-brown cap fading to dull yellow-grey. Shaggy margin to cap. Gills adnexed, crowded, whitish to dingy brown. Ring high on a slender fibrous stalk. Spore print dark brown.

Although harmless and sometimes considered edible it is not of high quality and **should be avoided** for fear of confusion with other rather nondescript poisonous mushrooms. Flesh pale, with a smell of new-ground meal.

Common and widespread in grassy copses and thickets, lawns.

A. dura: cap cracking; **Field Mushroom** and other *Agaricus* spp: free gills; **Poison Pie**: slimy caps.

April–July

SHAGGY INK-CAP OR LAWYER'S WIG

Cap to 15cm:6in high; stalk to 10cm:4in

Coprinus comatus

Shaggy white young cap eventually dissolves into an inky fluid as gills autodigest from the edge of cap inwards. Crowded whitish gills turn black from tips. Black spores.

Edible when all white. **Common Ink-cap** (*C. atramentarius*) **(1)** has a smoother grey cap, brownish in centre. Edible, but not with, or for a day or two after or before taking alcohol, when it **causes nasty symptoms** similar to the drug Antabuse.

Both are common and widespread, in clusters on pastures, lawns, by the roadside, by paths and on bare soil.

Many, but not all, *Coprinus* species have gills that turn black and dissolve; **Mica Ink-cap**; **Field Mushroom** and other *Agaricus* spp, gills do not dissolve; **Parasol Mushroom** and **Shaggy Parasol**: gills do not dissolve.

April–November

Volvariella speciosa

A tall elegant mushroom whose silky white stalk has a sac-like volva at base but no ring. Cap whitish to grey-brown. Gills pale, becoming salmon-pink. Spore print pink.

Although edible it is **not recommended**. The young mushrooms are completely enclosed in a white veil forming "eggs" like those of amanitas. The sticky cap is at first conical and then opens out.

Quite common and widespread in grass in broadleaved woods, on manured soil, gardens compost heaps, pastures on rich soil.

V. bombycina: shaggy or felty cap, grows on stumps or logs, especially of elm; **Amanitas**: ring as well as volva, white gills and spores; **Field Mushroom** and other *Agaricus* species: ring on stalk, no volva, purple-brown spore print.

June–October

These delicious wild cousins of the commercial mushroom are common in pastures and other grassland in early autumn. The genus *Agaricus* has white or brownish caps, a ring on stalk (but no volva), pinkish or greyish gills often becoming black, and a dark brown spore print.

Field Mushroom (*Agaricus campestris*) (**1**): cap 3–10cm:1¼–4in, silky white when young. Ring narrow. Deep pink gills become brownish. This tasty mushroom can be gathered in quantity from pastures in early autumn. *A. bitorquis*, two rings, edible. *A. macrosporus*: massive, cap to 30cm:12in, flesh flushes pinkish at base.

A. bisporus (**2**) is the wild form of the cultivated mushroom. It has a wider ring and pale fawn cap.

Horse Mushroom (*Agaricus arvensis*) (**3**): cap 8–20cm:3¼–8in, creamy then russet. All parts stain yellow. Gills pinkish grey to dull brown. Fleshy ring on stalk. **Do not mistake** this good edible mushroom for the **Yellow-staining Mushroom.**

YELLOW-STAINING MUSHROOM
Cap 6–15cm:2½–6in; stalk to 15cm:6in

Agaricus xanthodermus

Flesh at base of stalk stains deep golden yellow when this mushroom is cut lengthways. The silky white cap also turns yellow at edges and where damaged.

This is **highly indigestible** and causes severe gastric upsets if eaten. The bright yellow flesh at the base of the stalk is the main distinguishing feature. Like all *Agaricus* species the spore print is dark purplish brown.

Quite common and widespread in parks, gardens, roadsides and under broadleaved trees in grassy places.

Several good edible mushrooms also stain yellow or become yellow with age, such as *A. silvicola* in woods, and the **Horse Mushroom** in fields, but no other has the golden yellow flesh at base of stalk.

July–October

Macrolepiota (Lepiota) procera

This excellent edible mushroom has a thick double-edged ring which can be slid up and down the tall slender stalk, and a scaly parasol-shaped cap with shaggy edges. Gills cream.

The brown cap surface breaks up into a central brown patch surrounded by brownish beige scales on a paler ground. The scaly stalk has brown zones and a slightly swollen base. The pale flesh does not redden on cutting. Spore print white.

Common and widespread in pastures (sometimes forming fairy-rings), roadside verges, under hedges, and in grassy open woodland.

Shaggy Parasol and lookalikes; **Lepiota**: ring fixed, generally smaller, **avoid, some are deadly**; *Leucoagaricus bresadolae*: in clusters on sawdust or compost, **toxic**; **Field Mushroom** and other *Agaricus* spp: brown spore print.

July–October

97

Macrolepiota (Lepiota) rhacodes

Pale beige cap with coarse shaggy scales, greyish, beige, light or dark brown, and a smooth central area. Thick double ring moves on smooth stalk. Gills and spore print white.

The white flesh turns saffron yellow when cut or bruised and the gills also discolour on bruising. The creamy white stalk has a swollen base and bruises brownish. A good edible fungus.

Fairly common and widespread, in disturbed soil, on compost heaps, in grass, and under conifers.

The similar *M. venenata* has dark brown scales, and a ring in a single piece, **toxic**; **Parasol Mushroom**; **Shaggy Ink-cap**; **Field Mushroom** and other *Agaricus* spp: dark brown spore print.

July–November

Lepiota helveola

Pinkish brown flattish conical cap with rings of concentric scales on a paler ground. Gills pinkish. Fragile ring on a slender stalk. Spore print white.

This typical small *Lepiota* is one of several similar species which are **deadly poisonous**. The stalks are smooth, generally ringed with irregular zones of brown. **Do not confuse** them with the larger **Parasol Mushroom**.

This and similar species are rare. Deadly Lepiota grows in urban copses in warm places, occasionally found in southeastern England. Others are found in various habitats.

L. brunneoincarnata: more purplish in colour, no distinct ring, grassy copses, **deadly**; *L. lilacea*: pale purplish brown cap, in gardens, **toxic**; *L. pseudohelveola*: finer scales on cap, under broadleaved trees, **toxic**.

July–October

Lycoperdon perlatum

Club-shaped thin-walled white fruitbody is firm at first, then soft, puffing out a cloud of powdery spores if pressed. Outer surface covered with small blunt spines.

In this common puffball the spore mass becomes olive-green at maturity, and the fruitbody discolours to a dirty yellowish brown and loses its spines. Edible when still white and firm throughout.

Very common and widespread on the ground in pastures, parks, and woods.

Pear-shaped Puffball: on wood; *Calvatia excipuliformis*: larger, spore-filled head on stout stalk; *L. echinatum*: spines 3–4mm:⅛in; **Common Puffball**; *Bovista plumbea*: surface flakes off exposing grey inner wall; **Earthballs**: hard.

July–November

100

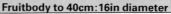
Lagermannia gigantea

1

This enormous puffball can weigh several kilos. It has a slightly flaky white outer surface. The internal spore mass is white becoming brown and powdery.

Good to eat when still white and firm throughout. **Common Puffball** (*Vascellum pratense*) (**1**) is smaller (3–6cm:1¼–2½in) and the spore mass is delimited from the sterile spongy base by a distinct membrane. Edible when young.

The Giant Puffball is locally common, in fertile pastures, gardens, hedgerows. The Common Puffball is very common and widespread in lawns, pastureland and grassy sand-dunes.

Other **Puffballs**; *Bovista plumbea*: white outer surface flakes off leaving bluish grey inner wall; **Earthballs**: hard and leathery throughout.

August–September; July–November (**1**)

CUP FUNGI
Fruitbody 6–8cm:2½–3¼in diameter

Peziza vesiculosa

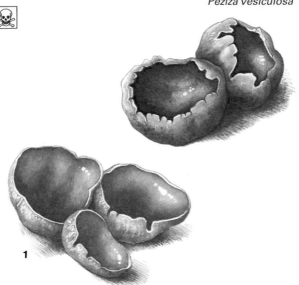

1

The pale brown fruitbody is at first round, becoming cup-shaped with an incurved edge. The inner surface (which bears the spores) is yellowish, the outer surface fawn.

It is **best not to eat** this or similar fungi. Although some are edible after cooking, others are **highly poisonous**. The cups of *Aleuria aurantia* (**1**), the **Orange-peel Fungus**, are 6–12cm:2½–4¾in across, and downy on the pale outer surface. Edible.

P. vesiculosa is rather uncommon, growing on manure heaps, rich soil, etc. The Orange-peel Fungus is more common, growing on paths, bare gravel and bare soil.

There are many cup fungi, often brightly coloured. *P. succosa*: bright yellow juice from cut flesh, in woods; *P. badia*: dark olive inner surface; *Scutellinia scutellata*: orange inner surface fringed with black hairs.

August–April; September–January (**1**)

Hericium clathroides (coralloides)

The brittle, coral-like fruitbody bears numerous long hanging spines 6–10mm: ¼–½in long, whose surface carry the spores. Creamy to ivory white at first; tough and pinkish with age.

This striking fungus is edible when young. The flesh is brittle and white but becomes tough in older specimens, although the spines themselves often remain edible and good.

Rather uncommon, on dead trunks of fir and of beech and other broadleaved trees.

The similar *Creolophus cirrhatus* consists of a series of flat white caps arising from a common base and bearing yellowish hanging spines on the lower surfaces. Not worth eating.

October–November

103

Ganoderma applanatum

Growing as a series of "brackets", the hard grey-brown bumpy upper surface becomes covered in a layer of rusty spores from the cap above. White undersurface turns brown when scratched.

The upper surface is shiny before it becomes dusty with spores. The **Birch Polypore** (*Piptoporus betulinus*) (**1**) (10–15cm:4–6in across) has a smooth matt greyish brown upper surface, white pores and flesh underneath. Spores white.

Ganoderma: on dead trunks of broadleaved trees and also on living trees, causing heartrot. Birch Polypore: on birch, causing heartrot, also on dead birches, common.

G. lucidum: shiny reddish top, stalked, rare; **Red-belted Polypore**; *Daedalea quercina*: corky grey-brown top, brown below, broad pores; *Heterobasidion annosum*: thin cap, top dark brown, white below, often at base of conifers.

All year round

POOR MAN'S BEEFSTEAK

Fistulina hepatica

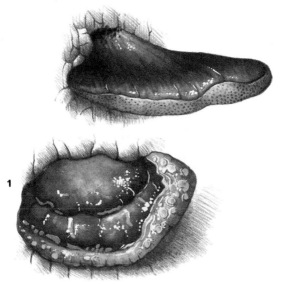

The cut flesh of this leathery topped shelf-fungus yields a red juice and looks and feels like red meat.

Rough reddish brown top, yellowish tubes on underside. Although edible after boiling, it does not live up to its name. **Red-belted Polypore** (*Fomitopsis pinicola*) (**1**) is a hard-topped bracket, reddish brown above, white below.

Poor man's Beefsteak: quite common on living oaks, causing "brown oak" staining in the wood; Red-belted Polypore; rarer, mainly on conifers, causing heartrot.

Phellinus igniarius: furrowed and cracked hard upper surface is grey to black with a velvety margin, underside yellow to brown-grey, on broadleaved trees.

August–November; all year (**1**)

105

Polyporus squamosus

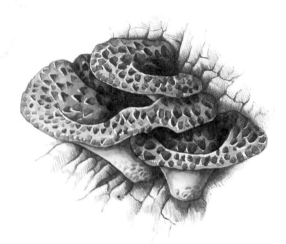

Circular to kidney-shaped polypore with a lateral stalk and a soft leathery surface breaking up into rings of pale to dark brown scales on a yellowish ground. White pores.

This is edible when very young and tender, but soon becomes tough. The white flesh has a faint smell of honey.

Very common on trunks, stumps and logs of broadleaved trees.

Polyporus species and relatives have soft leathery tops in contrast to the woody bracket-fungi. *P. varius*: smooth brownish yellow cap; the multiple-capped *Grifola frondosa* is reported to smell of "mice, hops and cold mashed potato".

April–December

Trametes versicolor

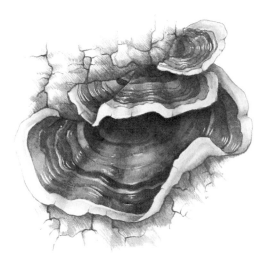

Thin silky-topped lobed-edged brackets are concentrically zoned in shades of olive, purple, green, grey and black. Underside whitish.

Logs are sometimes covered almost completely in the brackets of this common species. It is one of numerous small thin bracket-fungi that attack dead wood or sometimes cause disease in living trees.

Ubiquitous on dead wood of all sorts.

T. gibbosa: top white and felty, often with green growth of algae in centre; *Trichaptum abietinum*: thin elastic caps, velvety greyish top with violet tinge at edges, on softwood; *Daedalopsis confragosa*: pale brown zoned caps, white below.

All year round, but annual only

OYSTER FUNGUS
Cap 8–15cm:3¼–6in across

Pleurotus ostreatus

Fleshy semicircular or shell-shaped caps growing directly out of wood or attached with a very short (1cm:½in) stalk. Upper surface bluish grey fading to brown, white gills below.

A good edible fungus when young, with firm white flesh and a mushroom smell. Cap has a waxy sheen when young and colour can vary from almost black to dark brown in older specimens. Gills decurrent. Spore print lilac.

Common on dead trunks and branches of all kinds. Can also attack living broadleaved trees, and more rarely conifers.

Other common (but **not edible**) bracket fungi with gills are *P. cornucopiae*: whitish to beige, becoming funnel-shaped; *P. pulmonarius*: thin white brackets with lobed edges; *P. dryinus*: white scaly cap, cylindrical stalk.

All year round

108

VELVET FOOT

Cap 5–10cm:2–4in; stalk to 6cm:2½in

Flammulina velutipes

1

Clusters of flat, sticky, honey-coloured to bright red-brown caps on dark brown velvety stalks. Gills yellowish, adnexed. Spore print white.

The caps are edible after the slimy skin is removed but are not of high quality. **Sulphur Tuft** (*Hypholoma fasciculare*) **(1)** has a yellow cap (4–8cm:1½–3¼in), sulphur-yellow gills turning black, yellow stalk and brown spore print. **Inedible**.

Both are very common and widespread, growing on stumps and dead wood of broadleaved trees (Sulphur Tuft also on conifers).

Honey Fungus; *Pholiota alnicola*: narrow ring on stalk, brown spores, on alder and birch wood; other relatively common yellow pholiotas on wood have scaly caps, e.g. *P. squarrosa*.

September–March

109

Mycena galericulata

Conical pale grey-brown cap flattens out on aging. Gills white becoming pinkish. Stalk tough, smooth, same colour as cap, "rootlike" at base. Spore print white.

Inedible. Clustered Psathyrella (*Psathyrella hydrophila*) (**1**) has a moist chestnut brown cap drying to pale yellow in centre. Unstriated shaggy edge. Faint ring on white stalk. Gills adnate, greyish brown. Spore print black.

Both are common and widespread, growing in clusters on stumps of broadleaved trees.

Deadly Galerina and lookalikes; *M. polygramma*: pale greyish caps, grey stalk with fine greyish lines, in tufts; *M. inclinata*: reddish brown cap and stalk, rancid smell, in clusters; *Ps. candolleana* also grows on wood.

May–November; August–December (**1**)

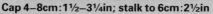
Galerina unicolor (marginata)

Sticky date-brown cap with darker striated edge soon expands and flattens out, drying a dull yellow-brown. Small ring on stalk, which is darker below ring. Spore print brown.

The narrow adnate gills are crowded, and are at first yellowish, becoming pale russet. The stem is fibrous. This is a **highly poisonous** mushroom and should not be eaten.

Common and widespread, growing in tufts on conifer stumps and other woody debris.

Kuehneromyces mutabilis: very similar, but stem sheathed in a woolly layer up to ring, on broadleaved stumps, not toxic; **Clustered Psathyrella**; *pholiotas* ring more distinct, scaly stalk; **Honey Fungus**.

September–November

MICA INK-CAP
Cap 2–4cm:¾–1½in; stalk to 8cm:3¼in

Coprinus micaceus

Fragile, bell-shaped, tawny brown cap is grooved from edge almost to centre and the central surface is covered with minute glistening granules which soon disappear.

Gills fragile, crowded, brown to black, eventually dissolving but not to the same extent as the Ink-caps. The edge of the cap is somewhat lobed and often splits as the mushroom ages. Stalk white with glistening granules near base. Spores black.

Very common everywhere on stumps and logs. When growing on ground it is in contact with buried wood or dead roots. Usually in clusters.

C. domesticus: very similar, granules on the slightly paler cap are more prominent and last longer; *C. disseminatus*: cap 0.5–1cm:¼–½in, pale brown or grey, velvety, fragile, grooved, gills not dissolving; **psathyrellas**.

All year round, especially May–December

112

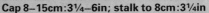

Tricholomopsis rutilans

Golden yellow cap speckled with minute purplish scales, especially in the centre. The adnexed, crowded gills are orange-yellow. Stalk yellow. Spore print white.

Although harmless this common mushroom is not worth eating. The yellow flesh smells faintly of damp wood and has a slightly bitter taste. The yellow cap becomes a rusty brown with age.

Common and widespread on and around conifer stumps, fence posts, etc.

Tr. decora: cap golden-yellow, only slightly speckled with olive; **tricholomas**: on ground; **Honey Fungus**: ring on stalk, brown scales on cap; *A. tabescens*: like **Honey Fungus** but no ring, in dense tufts on stumps, e.g. oak.

August–November

Omphalotus olearius

A bright orange fungus, becoming duller and darker with age.
Cap soon flat or sunken with a central knob. Gills narrow,
crowded, sharp-edged, decurrent. Pale cream spore print.

This **poisonous** fungus superficially resembles a larger version of
the Chanterelle. The smooth cap eventually turns brownish.
Edges of cap incurved.

Growing in clusters on wood. Very rare in Britain, only in
southeastern England. More common in southern Europe, Sweet
Chestnut being one of its preferred hosts.

Chanterelle: grows on ground; **False Chanterelle**; grows on
ground or on woody debris.

July–November

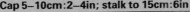

Collybia fusipes

Tough, flattened, grooved, spindle-shaped stalk supports a shiny reddish brown cap. Gills thick, greyish becoming speckled with brown, far apart. Spore print white.

The swollen stalks, brown at base, paler above, arise from a common base within the wood on which they grow. The cap is often rather irregularly shaped and sometimes develops rusty spots. The young caps only are edible after cooking.

Common and widespread, growing in tufts on stumps or arising from roots, especially of beech and oak.

Collybia confluens: on the ground, smaller, hairy pinkish stalks, **inedible**; **Russet Shank**. Collybias have tough stalks and white spores.

July–November

Armillaria mellea

1

Flattish honey-coloured cap with an incurved edge is covered with fine brown scales in centre. Gills white, crowded, adnate or decurrent. Large ring on stalk. Spores white.

It spreads by thick black "bootlaces" (rhizomorphs) running under the bark of trees it infects. Edible **with care. Orange Pholiota** (*Gymnopilus spectabilis*) (**1**) cap 8–15cm:3¼–6in, velvety, rich orange-brown, stalk thick, fibrous.

Honey Fungus: very common in tufts on stumps and above roots, a serious parasite of many trees; Orange Pholiota: less common, clustered at base of broadleaved trees or on stumps.

A. tabescens: similar to Honey Fungus, no ring, more southerly distribution; other pholiotas: scaly stalks, often scaly caps, spore print brown.

June–December

ROUND-STALKED AGROCYBE
Cap 4–12cm:1½–4¾in; stalk to 12cm:4¾in

Agrocybe aegerita

Smooth silky beige cap with yellowish centre becomes wrinkled and cracked. Gills adnate-decurrent, pale, becoming dull brown. Ring on stalk soon becomes shrivelled. Spores brown.

Pale flesh has a pleasant smell. Edible. Stalk white, greyish at base, covered with small fibres. **Slimy Beech Cap** (*Oudemansiella mucida*) (**1**) has slimy translucent white cap and gills. Spores white. Edible after cooking.

Round-stalked Agrocybe: in clusters on stumps or dead wood of many broadleaved trees; Slimy Beech Cap: only on dead trunks or branches of beech.

Honey Fungus; *Kuehneromyces mutabilis*: two-toned brown cap, which changes colour on drying, no ring on stalk.

May–November; August–November (**1**)

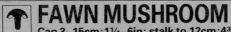

Pluteus cervinus (atricapillus)

Greyish brown cap with darker striations radiating from centre.
Gills crowded, free, white becoming pinkish. Stalk white, with
brown fibres. Spore print pink.

The stalk is slightly thickened at the base. The cap is bell-shaped
at first, then opening out with a central knob. Flesh pale and
smelling faintly of radishes. Edible but poor.

Very common and widespread on fallen trunks, stumps, etc.,
especially of broadleaved trees, and on piles of sawdust.

Other *Pluteus* spp. also grow on wood. Cap surface scaly, fibrous,
velvety or shaggy, stems fibrous, gills white to pinkish, free, spore
print pink. Some entolomas are very similar, but grow on the
ground and have attached gills.

All year round

Lycoperdon pyriforme

Beige or yellowish pear-shaped rough-surfaced fruitbodies grow in clusters. Apex slightly pointed with a central pore through which the spores are released.

This is the only puffball that grows on wood. The interior spore mass is white, becoming yellowish and powdery as the spores mature.

Common and widespread on stumps, fallen trunks and buried wood, usually of broadleaved trees.

L. lividum is similar but rounder and grows in grass; **Pearl-studded Puffball**: **Earthballs** are hard and leathery throughout.

August–November

JUDAS'S EAR
Fruitbody 5–12cm:2–4¾in across

Auricularia auriculae-judae

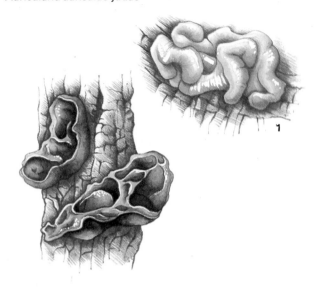

Ear-shaped greyish brown to reddish brown fruitbody has a velvety outer surface and firm jelly-like flesh when young and/or moistened. The inner surface bears vein-like ridges.

The whole fruitbody becomes more irregularly shaped with age and is bone-hard when completely dry. It is a good edible fungus. **Witches Butter** (*Tremella mesenterica*) (**1**) is golden orange, and of a softer consistency when moist, shrivelled and hard when dry.

Judas's Ear: common on dead branches of elder, occurring more rarely on other broadleaved trees. Witches Butter: common and widespread on dead branches.

A. mesenterica: irregularly lobed caps, whitish shaggy upper surface, on logs and stumps of elm; *Exidia truncata*: small, blackish, rounded, firm gelatinous flesh, minute warts on upper surface, especially on dead wood of oak.

All year round especially October–December

Daldinia concentrica

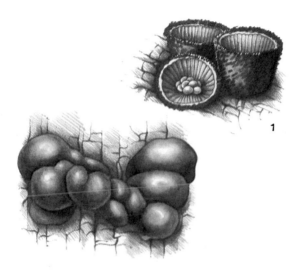

1

Hard, rounded, persistent fruitbodies, with a dull surface, at first dark brown becoming black, resembling burnt buns. When cut open, they show concentric silvery rings.

This fungus often appears on dead branches after fires. The **Bird's Nest Fungus** (*Cyathus striatus*) (**1**) 0.5–1cm: ¼–½in across, has a cup-shaped fruitbody with a fluted inner surface containing 10–12 small grey "eggs" which contain the spores.

Carbon Balls: common on dead branches of broadleaved trees, especially Ash. Bird's Nest Fungus: quite common on fallen branches and rotten wood.

Exidia glandulosa is smaller than Carbon Balls, with a firm jelly-like consistency. Of the other Bird's Nest Fungi, *C. olla* has a smooth inner face to "nest", *Crucibulum laeve* has a more cylindrical "nest" and creamy yellow "eggs".

All year round; March–November (**1**)

Index and Checklist

Keep a record of your sightings by inserting a tick in the box